Changes to Earth's Surface

Copyright © by Harcourt, Inc.

All rights reserved. No part of this publication may be reproduced or transmitted in any form or by any means, electronic or mechanical, including photocopy, recording, or any information storage and retrieval system, without permission in writing from the publisher.

Requests for permission to make copies of any part of the work should be addressed to School Permissions and Copyrights, Harcourt, Inc., 6277 Sea Harbor Drive, Orlando, Florida 32887-6777. Fax: 407-345-2418.

HARCOURT and the Harcourt Logo are trademarks of Harcourt, Inc., registered in the United States of America and/or other jurisdictions.

Printed in the United States of America

ISBN-13: 978-0-15-362077-5

ISBN-10: 0-15-362077-3

1 2 3 4 5 6 7 8 9 10 179 16 15 14 13 12 11 10 09 08 07

Visit *The Learning Site!*
www.harcourtschool.com

Lesson 1: How Does Earth's Surface Change?

VOCABULARY

crust
mantle
core
weathering
erosion
deposition
glacier

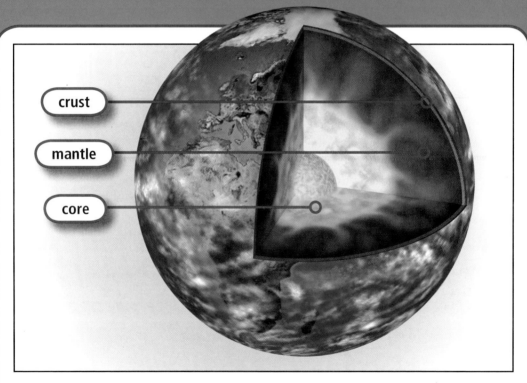

- crust
- mantle
- core

The thin, outer layer of Earth is called its **crust**.

The **mantle** is a very thick layer of Earth. It is found under the crust.

The **core** is found beneath the mantle. It goes to the center of Earth.

In **weathering**, something is broken down into smaller and smaller pieces. This rock used to be much bigger.

In **erosion**, the weathered pieces are removed. They travel to another place. Waves carried pieces of rock away. This made the cave.

In **deposition**, eroded material is dropped. The wind deposited sand to form these dunes.

Glaciers are huge sheets of ice. They stay frozen year-round.

READING FOCUS SKILL
MAIN IDEA AND DETAILS
The main idea is what the text is mostly about. Details are pieces of information about the main idea.
Look for information about what causes Earth's surface to change and details about each cause.

Earth's Layers

Have you ever cut an onion in half? You can see it has a thin outer layer. Inside, you can see more layers. Earth is like an onion. It also has a thin outer layer. Inside, it also has layers.

Earth's thin, outer layer is its **crust**. Dry land and the ocean floor make up the crust. Most of the living things on Earth live here. The crust is the only layer of Earth that people have seen.

Look at the picture below. Find the crust. Look at how thin it is compared with the other layers.

Earth's Layers

Within Earth's main layers lie several other layers that differ in their physical properties. The layers differ in temperature, pressure, and composition.

Outer Core (2200 km thick)
The outer core is made of very hot liquid iron and nickel. The movement of the outer core produces Earth's magnetic field.

Inner Core (1250 km thick)
The inner core is very hot, but it's been pressed into a solid metal ball by pressure from the layers around it.

The layer below the thin crust is the **mantle**. The mantle is a very thick layer. Like the crust, the upper mantle is solid. Beneath the solid upper mantle is a thick zone of hot, soft rock.

The layer below the mantle is the **core**. The core spreads from the mantle to Earth's center. Scientists divide it into the outer core and the inner core.

The outer core is made of a very hot liquid metal. The inner core is also very hot. But pressure has made the inner core a solid ball.

Scientists think the core is made of iron and nickel. Iron is magnetic. The iron in the core produces a magnetic field around Earth. Earth has a magnetic north pole and a magnetic south pole. A compass works because of Earth's magnetic field. The compass needle always points toward Earth's magnetic north pole.

 Describe Earth's layers.

Crust (0–100 km thick)
Earth's crust is the thinnest of Earth's layers. It is thinner under the oceans than under the continents.

Mantle (2900 km thick)
The mantle consists of two layers. The upper layer is the lithosphere (LITH•uh•sfir), with the asthenosphere (as•THEN•uh•sfir) lying below it.

5

Wind, Water, and Gravity Change Earth's Surface

Earth's crust is always changing. One way things change is by being broken down. Picture a large mountain. On a windy day, wind keeps hitting the mountain. This wind carries small pieces of sand and dust. When the sand and dust hit the mountain at a high speed, they break off tiny bits of rock. The mountain is undergoing weathering. **Weathering** is the process of being broken down into smaller and smaller pieces.

Water can also weather Earth's surface. If it rains on the mountain, the raindrops striking the mountain carry away small pieces of rock. This process is called erosion. **Erosion** is the removal and transportation of weathered material.

Gravity also erodes materials. Think about the mountain again. Gravity could cause a boulder on the mountain to roll down. Gravity can also cause rock slides and mudslides. During a rock slide, rocks bump into one another and break into smaller pieces. This is an example of both weathering and erosion.

Ship Rock is a landform in New Mexico. It was formed about 30 million years ago. Much of it has been worn away.

Weathering and erosion depend on many factors. For example, soft rock will erode faster than hard rock. Stronger forces also cause more weathering. A strong wind erodes more than a gentle breeze does.

Pieces of rock that are weathered and eroded are called *sediment*. Sediment is carried by wind and water. When the wind and water slow down, the sediment drops out. This is deposition. **Deposition** is the dropping or settling of eroded material.

Deposition can happen very close to where the sediment was produced. This often happens when gravity is the weathering agent. Deposition can also happen far away. Sediment picked up at the beginning of the Nile River in Africa may travel more than 6,600 kilometers (4,100 miles) to the river's mouth.

 What are three things that cause weathering?

The dunes are formed by wind deposition. ▼

The bends in the river were caused by deposition of sediment in the river.

Ice Changes Earth's Surface

You learned that liquid water changes Earth's surface. Frozen water also changes Earth's surface. In some parts of the world, it is so cold that huge sheets of ice form and stay frozen all year. These huge sheets of ice are called **glaciers**. Glaciers move very slowly as rivers of ice. Smaller glaciers are about the size of a football field. The largest glacier in the world covers the entire continent of Antarctica. This glacier is more that 4,000 meters (13,000 feet) thick. It is about one and a half times the size of the United States.

A glacier carries rocks, rubble, and sediment with it as it flows forward. It erodes small hills and other land areas. This smooths out the landscape. When the glacier stops moving or starts moving backward, it leaves behind large piles of sediment. This sediment can form an island. Long Island, New York, was formed this way.

A very heavy glacier can push down the land it flows over. When the glacier retreats, the land rises back up. This is called *uplift*. This process can cause cliffs to form.

 How do glaciers change Earth's surface when they flow forward?

A million years ago, small glaciers made these "hanging" valleys in Yosemite National Park.

Impacts Change Earth's Surface

Have you noticed that the moon's surface looks dented? Those dents are called *craters*. Craters form when meteorites strike the moon. *Meteorites* are rocks from space. Meteorites have hit Earth also. Fewer meteorites have hit Earth because Earth's atmosphere is thicker than the moon's atmosphere. Our atmosphere causes falling objects to burn up before they reach Earth's surface. But a meteorite can be large enough to make it through Earth's atmosphere. It may form a crater when it hits Earth's surface.

 Why does Earth's surface have fewer craters than the moon's surface?

This crater is in Arizona. It was formed by a meteorite that hit Earth about 20,000 years ago. It is about 1.2 kilometers (0.7 miles) across.

Review

Complete this main idea statement.

1. The three layers of Earth are the _____ , the _____ , and the _____ .

Complete these detail statements.

2. The wind breaking a boulder into smaller pieces is an example of _____ .

3. A river carrying away weathered material is an example of _____ .

4. Huge masses of ice that change Earth's surface as they flow are _____ .

Lesson 2

VOCABULARY
plate tectonics
mid-ocean ridge
sea-floor spreading

What Are Plates and How Do They Move?

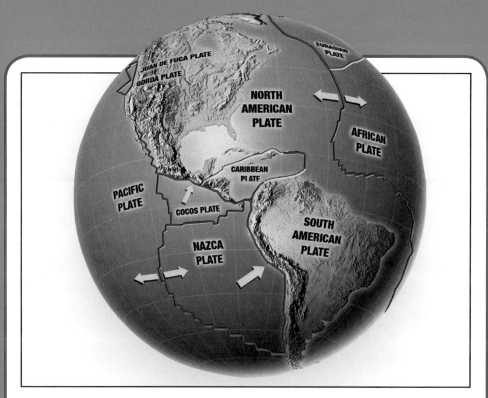

Plate tectonics is a theory about the upper part of Earth. The upper part includes the crust and the upper part of the mantle. Plate tectonics says that this upper part is divided into plates that are always moving.

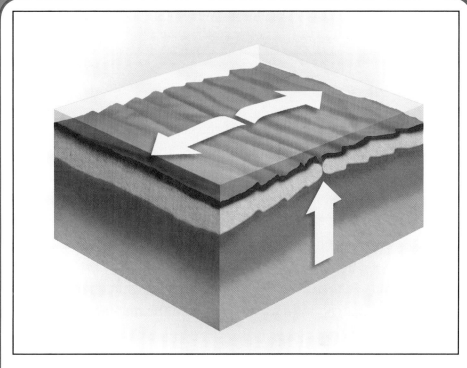

This diagram shows two plates moving apart from each other. Most plates move apart along a chain of mountains under the ocean. This chain is called the **mid-ocean ridge**. **Sea-floor spreading** takes place along the mid-ocean ridge. Liquid rock rises where the two plates move apart. This liquid rock becomes solid. The solid rock pushes the two plates farther and farther apart.

READING FOCUS SKILL

COMPARE AND CONTRAST

You compare and contrast when you look for ways things are similar and different. Compare means to find the way things are similar. Contrast is to find the way things are different.

Compare and contrast how Earth's plates move.

Earth's Plates

Plate tectonics is the theory that the crust and upper mantle are divided into plates that are always moving. The crust and the upper part of the mantle form the lithosphere. The map below shows how the lithosphere's plates fit against one another. They look like the pieces of a jigsaw puzzle. These plates are always moving.

The lithosphere moves very slowly. The plates move only a few centimeters each year. Over time, this causes major changes. Some plates may break apart. Others may move together and form one plate. New plates can form as hot rock from the mantle hardens.

 What is the lithosphere?

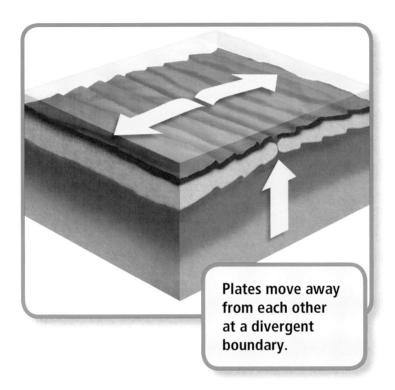

Plates move away from each other at a divergent boundary.

Plate Boundaries

Earth's plates meet at plate boundaries. There are three main types of plate boundaries: divergent boundaries, convergent boundaries, and transform fault boundaries.

At a divergent boundary, plates move away from each other. The picture above shows a divergent boundary. Most divergent boundaries are found along the mid-ocean ridge. The **mid-ocean ridge** is a chain of mountains that runs about 83,000 kilometers (52,000 miles) through the world's oceans. Along the highest part of the mid-ocean ridge is a rift. A **rift** is a deep valley where plates move apart. As the plates move apart, hot rock from the mantle moves up. The hot rock cools and hardens. This new solid rock pushes the sides of the boundary farther away from each other. This process is called **sea-floor spreading**.

Another type of plate boundary is a transform fault boundary. Here, plates move in opposite directions. They grind past each other. This movement can cause earthquakes.

At a convergent boundary, plates push into each other. There are three kinds.

In the first kind, two continental plates push into each other. The colliding plates fold and bend. This forms mountain ranges.

The second kind occurs where two oceanic plates push into each other. One plate is pushed down. This can form a deep-ocean trench.

The third kind occurs when a continental plate collides with an oceanic plate. The oceanic plate is denser. It is pushed under the continental plate. Mountains and volcanoes form along this boundary.

 Compare and contrast the three kinds of plate boundaries.

Plates move past each other at a transform fault boundary.

Plates move together at a convergent boundary.

Plate Movements Change Earth's Surface

Scientists have looked at different events to find out how the plates have moved over the years. They studied how the shapes of some continents, such as South America and Africa, seem to fit together. They found that the rocks and soil types along the edges of some continents are the same. They also see that the same kinds of plant and animal fossils are found on different continents, even though they are now far apart.

Scientists have concluded that all of the continents once formed one large continent. They call this continent *Pangea*. It existed about 220 million years ago. Pangea was not the first supercontinent. Scientists believe that the continents have joined together several times during Earth's history. They then pulled apart again.

Today the Atlantic Ocean is growing and the Pacific Ocean is shrinking. This is causing the Americas to move farther away from Europe and Africa and closer to Asia.

 Compare the methods scientists use to explore how Earth's plates have moved.

Complete these compare and contrast statements.

1. Plates move toward each other at a _____ boundary and away from each other at a _____ boundary.

2. A mountain can form at a _____ boundary, and an earthquake can happen at a _____ _____ boundary.

3. Today, the _____ Ocean is growing larger while the _____ Ocean is becoming smaller.

Lesson 3

What Causes Earthquakes and Volcanoes?

VOCABULARY
fault
earthquake
focus
epicenter
volcano

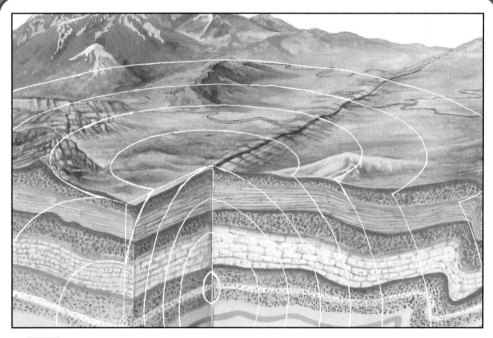

A **fault** is a break in Earth's crust where rocks can slide past each other. The point inside Earth where an earthquake begins is called the **focus**. The area on Earth's surface directly above the focus is the **epicenter**.

The picture shows earthquake damage in Loma Prieta, California. A major earthquake hit it in 1989. An **earthquake** is a vibration in Earth's crust. It happens when there is a release of energy at a fault.

A **volcano** is a mountain formed from hot or molten rock. The molten rock is pushed to Earth's surface and builds up.

READING FOCUS SKILL
CAUSE AND EFFECT

A cause is what makes something happen. An effect is what happens.

Look for causes and effects of earthquakes and volcanoes. Identify how plate tectonics are related to these events.

Earthquakes

Earth's plates are under pressure. The plates may bend. If the pressure is great enough, the rocks that make up the plates may break. When the rocks break, a fault is formed. A **fault** is a break in Earth's crust where rocks can slide past each other. Most faults form along plate boundaries. This is because the pressure at the boundaries is great. Faults may also form within a plate.

As Earth's plates move, pressure builds up along the faults. The rocks bend and stretch. If the rocks at the fault are bent or stretched too far, they will break. When they break, they release energy. This release of energy causes vibrations in Earth's crust called an **earthquake**.

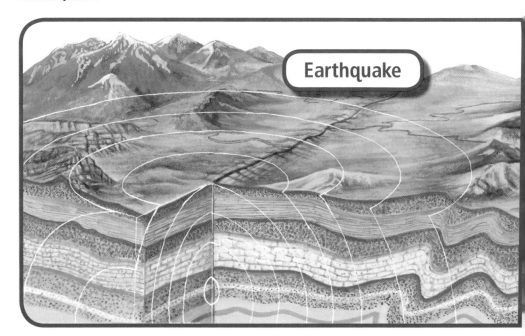

Earthquake

Earthquakes can occur close to Earth's surface or deep inside the crust. The point inside Earth where an earthquake begins is called the **focus**. The area on Earth's surface directly above the focus is the **epicenter**.

When an earthquake occurs, the released energy travels in waves away from the focus. The fastest waves are primary waves, or *P waves*. These waves are the first waves to be detected. Their movement is like the movement of an accordian.

The second-fastest waves are secondary waves, or *S waves*. S waves move across the direction in which the P waves are traveling. They may move up and down or side to side.

P waves and S waves move through the inside of Earth. Another kind of wave, a *surface wave*, moves only along Earth's surface. Some surface waves shake the ground. Other surface waves roll across the ground. Surface waves cause most of the damage done to buildings during an earthquake.

 What is the effect of rocks breaking at a fault?

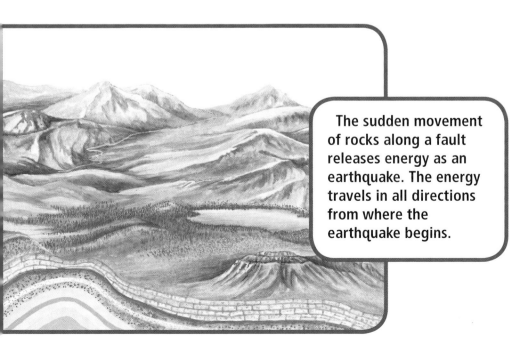

The sudden movement of rocks along a fault releases energy as an earthquake. The energy travels in all directions from where the earthquake begins.

19

Measuring Earthquake Strength and Damage

Scientists use a tool called a *seismograph* to measure and record the motion of an earthquake's energy waves. Seismographs can also find an earthquake's location.

In 1935, Charles Richter developed a scale of numbers based on seismograph records. His scale measures an earthquake's strength. The *Richter scale* estimates how much energy an earthquake releases.

Another scale also measures an earthquake's strength. The *moment magnitude scale* is not based on seismograph records. It is based on how far rocks move along the fault.

The *Mercalli intensity scale* measures how much damage an earthquake causes.

An underwater earthquake in the ocean can cause a giant wave. This is called a *tsunami*. A tsunami can travel long distances. When it comes to the coast, it can be 30 meters (100 feet) high. This giant wall of water slams into the land. It can damage buildings, cause erosion, and take all the sand away from a beach. An underwater volcano can also cause a tsunami.

 What causes a tsunami?

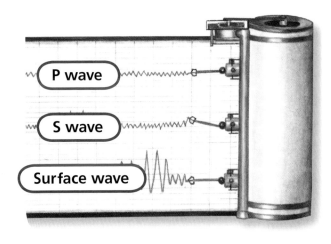

◀ This seismograph records earthquake waves. Note how the lines change when the Earth moves.

Volcanoes

You know that some mountains form when two continental plates meet. Some mountains form another way. Volcanic mountains form when molten rock is pushed to the surface and builds up. This molten rock is called *magma*.

Most volcanoes form along the boundaries of Earth's plates. Some volcanoes form at divergent boundaries. The plates pull apart. Magma rises to the surface. Other volcanoes form at convergent boundaries. When oceanic crust is pushed down, the crust sinks. As the crust sinks, some of the rock becomes hot enough to melt. This forms magma. The magma is lighter than the crust, so it is pushed up. If there is an opening to the surface, the magma can explode. When it explodes, it forms a volcano.

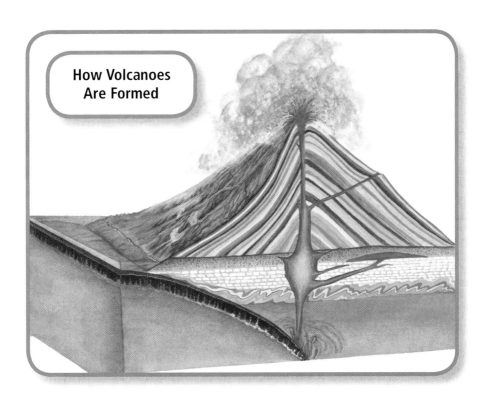

How Volcanoes Are Formed

There are three types of volcanoes: shield volcanoes, cinder cone volcanoes, and composite volcanoes. The type of volcano depends on the type of volcanic eruption that formed it.

Volcanoes can erupt explosively. Materials are thrown violently out of the vent. This forms cinder cone volcanoes.

Volcanoes can also erupt calmly. The lava slowly flows out of the vent. This forms shield volcanoes.

Composite volcanoes are the most common type. They are formed by a combination of explosive and nonexplosive eruptions.

 What causes a cinder cone volcano to form?

Shield volcanoes are broad and slightly dome-shaped. ▶

▲ **Cinder cone volcanoes have steep sides.**

▲ **Composite volcanoes are the most common.**

Hawaiian Volcanoes

Magma can occur under a plate away from a boundary. This area is called a hot spot. The magma is less dense than the solid rock. So the magma is pushed up toward the surface. If the magma erupts above the surface, it can form a volcano.

As the plate moves over the hot spot, the volcano stops erupting. A series of new volcanoes can then form. If the hot spot is under oceanic crust, the series of volcanoes can become a chain of islands. This is how the Hawaiian Islands formed.

 What causes a chain of volcanic islands to form?

The Hawaiian Islands formed when the Pacific plate moved over a hot spot. Magma flowed from inside the Earth and hardened.

Complete these cause and effect statements.

1. Rock breaking at a _____ can cause an earthquake.
2. An underwater earthquake or volcano may cause a large wave called a _____.
3. When magma is pushed to the surface and builds up, a _____ is formed.
4. A _____ volcano is formed by a mixture of explosive and nonexplosive eruptions.

GLOSSARY

core (KAWR) The layer of Earth that extends from Earth's center to the bottom of the mantle.

crust (KRUHST) The thin, outermost layer of Earth, which includes both dry land and the ocean floor.

deposition (dep•uh•ZISH•uhn) The dropping or settling of eroded materials.

earthquake (ERTH•kwayk) A vibration in Earth's crust, caused by the release of energy in a fault.

epicenter (EP•ih•sent•er) The area on Earth's surface directly above the focus of an earthquake.

erosion (uh•ROH•zhuhn) The removal and transportation of weathered materials.

fault (FAWLT) A break in Earth's crust where rocks can slide past each other.

focus (FOH•kuhs) The point inside Earth where an earthquake begins.

glacier (GLAY•sher) An immense sheet of ice that stays frozen year-round.

mantle (MAN•tuhl) The thick layer of Earth beneath the crust.

mid-ocean ridge (mid•OH•shuhn RIJ) A chain of mountains that runs through the world's oceans.

plate tectonics (PLAYT tek•TAHN•iks) The theory that the lithosphere is divided into plates that are always moving.

sea-floor spreading (SEE•flawr SPRED•ing) The process, which takes place along the mid-ocean ridge, in which liquid rock rises, becomes solid, and pushes two plates farther and farther apart.

volcano (vahl•KAY•noh) A mountain formed when molten rock is pushed to Earth's surface and builds up.

weathering (WETH•er•ing) The process of being broken down into smaller and smaller pieces.

Think About the Reading

1. What can you do to help you remember what you have learned in this chapter?
2. What questions do you have after reading this book? How can you find the answers to your questions?

Hands-On Activity

Use modeling clay to make models of the three types of tectonic plate boundaries.

1. First, model what happens at a divergent boundary. How can you show the new crust?
2. Model what happens at a transform fault boundary. How is it different from a divergent boundary?
3. Last, model what happens at a convergent boundary. Which convergent boundary did you model?

School-Home Connection

Tell a family member about the slow and rapid changes to Earth's surface that you learned about in this chapter. Work with your family member to make a chart. Show things that change Earth's surface quickly and things that change Earth's surface slowly. Discuss which ones occur around your home.

| GRADE 6 |
| BL Book 6 |
| WORD COUNT |
| 2025 |
| GENRE |
| Expository Nonfiction |
| LEVEL |
| See TG or go Online |

Harcourt Leveled Readers Online Database
www.eharcourtschool.com

ISBN-13: 978-0-15-362077-5
ISBN-10: 0-15-362077-3

Photo Credits: Cover: Digital Image 1996 Corbis/Corbis; p. 10: Harcourt; p. 11: Jerry Lodriguss/Photo Researchers, Unc.; p. 12: John Chumack/Photo Researchers, Inc.; p. 13: John Chumack/Photo Researchers, Inc.; p. 14: Nasa/Science Source/Photo Researchers, Inc.; p. 15: StockTrek/ Getty Images; p. 2: John Chumack/Photo Researchers, Inc.; p. 3: (tl) Celestial Image Co./Photo Researchers, Inc.; (tr) Noao/Photo Researchers, Inc.; p. 4: Nasa/Photo Researchers, Inc.; p. 5: Nasa/Science Source/Photo Researchers, Inc.; p. 6: Chris Cook/Photo Researchers, Inc.; p. 8: David Parker/Photo Researchers, Inc.

If you have received these materials as examination copies free of charge, Harcourt School Publishers retains title to the materials and they may not be resold. Resale of examination copies is strictly prohibited and is illegal.

Possession of this publication in print format does not entitle users to convert this publication, or any portion of it, into electronic format.